"Valuable tips and expert advice for working smart in the brave new world of business technology...Packed with essential, practical information for business newcomers but can also serve as a valuable resource for those with more experience."

Kathleen Hughes
*ALA Booklist*

"A must read for people trying to get ahead in today's fast-paced and complex world. Milrod and VanDam present their tactics, techniques and strategies in plain English. This book *will* help you succeed."

Eric Gelb
Author
*Personal Budget Planner*

"Full of helpful tidbits, *Mastering Communication Through Technology* provides simple how-to's and practical advice for those of us who are not technically inclined. I want all of my employees to read the chapters on e-mail etiquette! Thank you to Ms. Milrod and Mr. VanDam for this surprisingly useful guide."

Cara Rubinstein Hoxie
President and CEO
Rehabilitation Services of Northern California

"This book provides excellent advice for professionals communicating in today's high-tech business world. I highly recommend incorporating these principles in every company's employee handbook!"

Nancy Casey
CEO
Dpmed Inc.

"This book is easy to read and full of helpful hints. I have found it useful in my business and personal communications and will recommend it to my clients."

Rachel D. Bass
Founder and Principal
Zephyr Consulting

For more reviews and other information about this book, visit our Web site: http://www.smallbusinessadvice.com.

# Mastering Communication Through Technology

Copyright © 2001 by Eve Milrod and Arthur VanDam

Published by:

Career Advancement Center, Inc.
Woodmere, New York, USA 11598-0436

This book is published under the copyright laws of the United States. All rights reserved. No part of *Mastering Communication Through Technology* may be reproduced, stored in a retrieval system, or transmitted, in any form or by any means, electronic, mechanical, photocopying, recording, or otherwise, without express written permission from the Career Advancement Center, Inc.—the publisher, except for the inclusion of brief quotations in an article or review.

Book design: Mark Berman, makabe@scn.org

Front cover image © The Stock Market 2000

**DISCLAIMER.** This book, *Mastering Communication Through Technology*, contains advice about technology, communication, writing, public speaking and business. But the use of a book is not a substitute for technology, communication, publishing, public speaking, business, legal, tax, accounting, consulting or other professional services. Consult the appropriate professionals for answers to your specific questions. Neither the publisher nor the authors bear any liability for the incorrect or improper use of this book or the information and advice contained herein. If you do not wish to be bound by the terms of this paragraph, promptly return this book for a complete refund.

Career Advancement Center is a trademark of the Career Advancement Center, Inc. All other trademarks are the property of their respective owners.

ISBN: 1-890158-09-7

LCCN: 00-104760

Printed in the United States of America

# Table of Contents

Introduction .................................................................................. 1
Communication Strategy ........................................................... 3
Develop Your Own "To Do" List ............................................. 10
Get Organized—On-Line or Otherwise ................................. 11
Valuable Assets ......................................................................... 13
Letters and the Pony Express ................................................. 14
Business Cards .......................................................................... 15
Business Communication Across Cultures ......................... 18
Savvy E-mail Use ..................................................................... 20
Voice-mail Victory ................................................................... 24
E-mail and V-mail Responsiveness ....................................... 26
Security ...................................................................................... 27
The Paperless Society .............................................................. 29
Personal E-mail ........................................................................ 31
E-mail Do's and Don'ts ........................................................... 32
Staying in Touch ...................................................................... 33
Conference Calls ...................................................................... 34
Nothing Beats Face-to-face .................................................... 37
Pager Etiquette ......................................................................... 38
Legally Binding FAXes and E-Signatures ............................ 40
Some Helpful Wireless Phone Facts ..................................... 42
PDA'S and Electronic Organizers ......................................... 44
The Future is Here Now—Well Almost ............................... 46
One Final Word…For the Moment ....................................... 48

Other Books and Resources:
    How to Read Legal Documents (new publication) ..... 51
    The Career Advancement Center Catalog .................... 52

# The Authors

**Eve Milrod** currently lives in Silicon Valley, where she splits her time between working as a marketing executive in the software business, tinkering with her garden of flowers and hiking in the Santa Cruz Mountains. Before California, Eve lived in Tucson AZ, where she was principal of LeMill Associates, a business and marketing strategy consulting firm specializing in High Tech clients. Originally from New York, Eve worked for 7 years for a software division of Chemical Bank (now part of The Chase Manhattan Corporation) in a variety of marketing and operating positions. Eve holds an AB from Smith College and an MBA from Columbia University.

**Arthur VanDam** is responsible for the career development division of the Career Advancement Center. Arthur has extensive experience in career counseling, banking and finance, and accounting. He is a graduate of the Wharton School at the University of Pennsylvania. He enjoys playing soccer and the trombone.

# Dedication/Acknowledgements

## From Eve

To my family and friends, Who support me no matter which mountain—physical, virtual or spiritual—I attempt to climb.

## From Arthur

To my family. To my co-author Eve Milrod. And my editor and publisher at the Career Advancement Center.

· · · · · · · · · · · · · · · · · · · · · · · · · · · · · · · · · · · · ·

Special acknowledgement to Neeracha Taychakhoonavudh: thanks for sharing your insights about Asian customs.

To Mark Berman: thank you for your book design.

· · · · · · · · · · · · · · · · · · · · · · · · · · · · · · · · · · · · ·

# Foreword

Today, technology enables us to communicate with people across the globe instantaneously. You can send and receive voice messages and electronic or "e"-mails on your wireless phone or on your handheld wireless data device. Maybe you are reading this book in its electronic format—as an e-book. You can push the Reply icon in e-mail or on a pager; and you can FAX back a document.

Technology's "black box" combines many complex instructions behind the scenes and invites us to push one button or enter just a few keystrokes to execute an operation. The ease with which we can use these devices can make it seem that the impact of our words and actions is equally straightforward and nonchalant. But this is far from the truth. It is imperative for you to consider the consequences of your words and communiqués and to strategize and shape your speech and writing.

As people can find you all the time and work, family and other life pressures increase, it becomes increasingly important to use technology wisely and selectively—to your advantage. It has become crucial to *slow down and think carefully* before you act. As you read this book, develop a well-conceived communications strategy and create your own communication rules. Then follow those rules. Decide what message you want to send and who should receive your message. Decide what types of things you want to say. Consider the consequences of your actions and words, and proceed on *your* own timetable. Then you'll find that you can be even more effective in whatever activities and endeavors you pursue.

A few years ago, my employer decided that all the investment bankers should carry two-way pagers. Many of the bankers did not want to "get wired." But, as I waited in line to receive my pager, I thought about how I could use the pager as a way for my clients and colleagues to contact me. This technology could help me enhance my image as a "go to" person. In other words, my customers could find me more easily and I would be more available to help them.

The instructions to my pager explained how to use the device's features and how easy the device made communicating. As a banker, salesperson, writer and public speaker, I am very familiar with the power of words. It occurred to me that I should establish a communication strategy to help me use this new technology effectively, and more broadly, become a more effective businessperson.

The thought of carrying a pager reminded me of my Grandpa Max, a doctor. Whenever his patients needed care, they called his answering service. This communication link helped Grandpa help more patients and be a more effective doctor. But communication doesn't just involve e-mail, pagers, telephones and FAX machines. With technology, communication can be infinite. The key is to use each system to help you be more efficient and effective and to help you achieve your goals and attain greater success.

Eve Milrod and Arthur VanDam have created this valuable book to help you advance your career, your business and your personal life. Digest the pages and learn how to use technology wisely and formulate your own communication strategy.

Communicate wisely and more effectively.

Eric Gelb
Author and Publisher
*Personal Budget Planner* and
*Book Promotion Made Easy*

# Introduction

Technology is uniquely positioned in our world: it has the capacity to bring people much closer together at the same time it pushes us apart and adds huge complexity to our lives. With the development of technology—particularly the proliferation of wireless telephones and electronic mail ("e-mail") over the Internet, people nearly everywhere in the world can communicate regularly, easily and immediately. With this expansion of communication, there are many ways in which we have lost our privacy and our freedom. We have more opportunities in more mediums than ever to mis-communicate. These factors make it even more important to create and implement a communication strategy unique to your goals.

Today, most leading companies aggressively embrace Information Technology and advanced means of communicating, marketing and selling. These organizations are often fully "wired for sound," with e-mail, voice mail ("v-mail"), facsimile ("FAX"), wireless phones and pagers. In fact, you probably have access to or are using many of these technologies today. With all of this ever-expanding technology, it is important for you to find ways to use these technologies to your advantage, rather than permit the technologies to take control of your life. This book presents a broad array of communication strategies and tactics: it is designed to help you enhance your ability to prosper in this era of new technology and the new economy.

One fascinating aspect of our times is convergence. We see this in many places in business and society. For example, your wireless telephone can, or will soon, send and receive data and e-mail. This creates the convergence of voice and text. But perhaps the most dramatic shift is just beginning: the convergence of the "old economy" and the "new economy." You are seeing this with online purchasing and supply marketplaces. But this is just the beginning. To be more successful, you should continually seek out ways to implement technology in your business and daily life.

While providing some field-tested tips for using advanced communication media, this guide can also help you humanize your communications and communicate more effectively and directly with your audience to achieve your goals. Today more than ever, we all need to communicate with other people quickly and frequently, and it is essential to use our time wisely—this skill will become more critical with the upcoming technology advances. One of the recurring themes in our discussion will be about the need to *slow down* and make sure you are aware of and controlling your interactions and exchanges with others. Don't let the new technologies run your life and speed up your actions to the point where you act before you think. When you master technology, select the best options for you, and implement your choices consistently and effectively, you will find yourself in better communication with others, more productive, and more in control.

See you on the super highway.

Eve Milrod

C/O Career Advancement Center, Inc.
PO Box 436
Woodmere, NY 11598-0436 USA
E-mail: caradvctr@aol.com

# Communication Strategy

The key to success in any endeavor is establishing a plan or strategy. Communication is no exception to this rule. Imagine for a moment that you are about to make an important speech or presentation to one of your key customers, colleagues, or classmates. You walk up to the podium. First, you notice the 200 people who fill the auditorium; then you realize that you know these people and they have just become your potential critics. Next, you notice that there is a PC-compatible slide projector instead of the old fashioned carousel version you requested—the one that goes with your slides. What's more, your eyeglasses fog up, and you fumble to find your presentation notes. Not a comforting moment...but, with a communication strategy, you can be a more effective speaker, businessperson and leader.

There are two important components to an effective message: *the words and the delivery*. Both must be targeted to the audience. Think of your favorite actor or actress. If s/he is versatile, with each role the actor will bring out a different style and demeanor. In other words, the actor becomes the character s/he plays. This should be your goal— you want to play to your audience.

What you say and how you deliver your message will determine your level of success. You could craft the most cogent argument and wise plan, but if your tone is humorous while your audience is serious, you will fail. If your tone is angry, you are likely to alienate your audience and your message is likely to be ineffective. Or, you might deliver your message effectively, but the audience may have little interest in your topic or lack the authority to make a decision. In any of these situations, you are unlikely to succeed.

The key is to develop a communication strategy. This means setting goals and deciding what you want to say, how you want to say it and to whom you want to say it. You might read this last sentence and say, "This is impractical. When I am part of a conversation, I don't have time to pause and craft my words." There is merit to this view, but it is important to make sure that you say the right words.

**Audience Expectations**. Often, your audience has formed preconceived ideas about you and your presentation before you even begin. For example, you send a cover letter and resume to a company. Based on your material and your background, your interviewer has formed an opinion of you before you make eye contact. Your meeting will either substantiate or disprove that expectation. When you develop a communication strategy, it is essential that you consider your audience's expectations for you and what expectations you want to set with them. Then develop your message and tone accordingly. You can change the minds of the people in the audience, but it takes effort and often persistence.

**What image do you want to create?** Camille Lavington, executive consultant and author of the book *You've Only Got 3 Seconds to Make a First Impression* says, "People size you up in three seconds; they make a snap judgment about you and that's it. So, you have to think carefully to develop and cultivate your self-image in order to project the precise image that you want to project."

Your appearance and posture can have a dramatic effect on your image and the audience's perception of you. Choose your clothes and hairstyle carefully in order to project the image you want. When you stand up straight, you often convey a sense of confidence and authority. When you slouch and lean on the podium, the audience is likely to view you as unsure of yourself. When you smile, often you can typically disarm an audience, but a smile may be inappropriate for a serious meeting. You may dilute the value and power of your message, especially if you are delivering important news and information.

**Be polite**. Regardless of the communication medium you use, be polite. The authors of this book have observed many people being rude and condescending in business and personal situations. You will be much more successful in your life when you treat other people with kindness, respect and courtesy.

**Give a firm handshake**. When you meet someone, shake hands firmly and establish direct eye contact—look the other person in the eyes. These gestures demonstrate that you have self-confidence. Add a warm "hello" and a smile and you have a winning combination. While e-mail and the telephone cannot replace face-to-face (F2F) interaction, through your choice of words and tone you can establish a positive rapport with other people.

**Learn about your audience before you design and deliver your message**. When you address different people, you are likely to want to adapt your message and delivery to that audience and to meet your particular objectives. For example, you may adopt one style when you meet with your supervisor, another with your staff, another with clients, a fourth with friends, and yet another with your spouse or partner. And to complicate matters, the composition of the audience, the nature of your relationship, your message and your prior dealings will set the tone for your current and future interaction. For the most part, you can change the past, but it tends to be difficult, especially since people are quick to form a first impression.

**What do you want to communicate?** This may seem like a relatively simple and straightforward question. Yet, the speaker or the listener may garble the message. What you may understand in one way, your audience may interpret in a completely different fashion. Perhaps half your audience will interpret your message one way, one-quarter in another way, and the other quarter will wonder what you intended to say or what you meant. Your choice of words and phrases may seem clear and familiar to you, but those same words to an anxious staff member or a customer may have a different meaning. Or, they may have a different connotation that may leave the listener wondering what you really meant. It is critical to understand and guard confidential or potentially inflammatory information, and deliver this only to the intended audience. You will develop more savvy as you gain more experience in your field. Take time to plot out your message so your words are most likely to tell your audience what you want them to learn.

**Think about what you want to say.** Use clear and simple language. Use language that clearly explains what you mean. Use words, phrases and descriptions that are commonplace to your audience. In other words, avoid unnecessarily sophisticated words and phrases that may confuse and bore your audience. Consider obtaining a book that will help you build your vocabulary and improve your phraseology. Picture in your mind your audience's reaction to your message and delivery. Then practice your delivery to improve your communication style.

**With whom do you want to communicate?** Whom is your intended audience? With oral delivery, it can be complicated to select your audience. If you speak to a small group, usually you can select your audience. You can call a meeting or gather one or two people in your office or conference room. At a large gathering or party, audience selection may be more difficult. And oral communication doesn't always scale for large numbers: when you address a large group, it is wise to consider how much information you can realistically deliver and what the audience will absorb. Especially when addressing large groups, it is critical to shape your message accordingly.

Written communication has advantages and disadvantages.. First, you have more time than when speaking to compose and organize your thoughts. In most cases, you can edit your words. And to some extent you can manage who receives your message. Of course, people can share your letter with others or forward your e-mail infinitely. This is why when you communicate in writing you must manage both your message and your delivery very carefully. Written communication is a permanent record of your words.

**What is the best communication medium for your message?** Make it a practice to commit serious messages to paper—on paper or via electronic mail ("e-mail")—to memorialize your words. Or, you may want to deliver your words face-to-face ("F2F")—so you can have direct eye contact with the other person or people. Perhaps you don't want your comments in writing because you don't want a written record of your comments to circulate. Each situation will dictate a different course of action.

Different situations will require you to communicate orally or in writing. For better or worse, technology has led to the convergence of voice and text. You can use an optical scanner to convert written material into digital form and the reverse. You can convert voice into text and vice versa. Generally speaking, you should make a decision about how you want to send your message. Sometimes, writing helps you crystallize your thoughts and deliver a uniform message to large groups of people at the same time. For example, it is common today for executives in a company to compose an e-mail and distribute it to all employees at the same time.

While the medium and method of communication may have changed, the essence of communication remains the same. One person wants to deliver a message to one or more people or groups of people. When you want to communicate a particular message, it may be clear to you what you want to say and how you want to present your message. Will the recipient interpret the message in the way that you intended? It is important to note that the recipient may not comprehend your message or more likely, you and the recipient will interpret the same message differently. When your message is important, it is worthwhile to slow down to think about the meaning of your message and the best way to convey that message.

**Technology and communication.** As technology helps us communicate faster and faster, we need to pay greater attention to formulating a communication strategy and communicating more effectively. First, decide what you want to achieve. Perhaps you want to close a sale, be promoted, land a new position, establish a new relationship, or deliver bad news. Once you decide what you want to achieve and who your audience is, craft your message carefully and determine the best way to convey that message.

Today, technology plays an important role in the way you can deliver your message and the speed at which you can deliver it. As a result, it is important to utilize technology carefully and effectively. Take your time to make sure that you are sending the message you intend, to the intended audience, in the appropriate medium, on your own timetable.

**Establish your own set of communication rules**. Many people find that they are much more effective and consistent when they develop a set of rules. This applies to communication strategy. Consider creating your own set of rules or action steps that you will use to govern and guide your communication activities and actions. First, decide what you want to say and strategize before you communicate. Next, set down rules regarding what you will and will not say, write or reveal. Then most important, follow these rules consistently. Update your rules over time and as you evolve and face new circumstances and situations.

• • • • • • • • • • • • • • • • • • • • • • • • • • • • • • • • • • • • • • • • • • •
Communication success comes by creating a plan and implementing it consistently over time.
• • • • • • • • • • • • • • • • • • • • • • • • • • • • • • • • • • • • • • • • • • • •

# Develop *Your Own* "To Do" List

Today, the pace of work and the pace of change seem to continually accelerate. Bosses, partners, customers, schools, governments, friends and family seem to place greater and greater demands on our time and on our lives. Customer service professionals boast that they are available 24X7: twenty-four hours a day, every day. This total accessibility has its advantages, but also its costs.

Make it a practice to keep To Do Lists. You may use a pad, notebook, scrap paper clamped together, or an electronic organizer. Regardless of the system you choose, get in the habit of writing down all the things that you have to do, even if you assign a relatively low priority to some of the tasks. When you write down all items in a well-organized fashion and store your list in a convenient place, you can take comfort that you'll read the task list and remember about the To Do. Of course, this assumes that you review your To Do List on a regular basis.

**Prioritize your tasks**. Many people find it helpful to create a To Do List in connection with a daily calendar. This way, they can prioritize events and responsibilities by day. In other words, an item that is very important and has a short deadline might appear in your diary "today." Less pressing items will fall to the bottom of the list. When you use an electronic diary, software program or personal digital assistant ("PDA"), you can prioritize tasks and To Do's, copy events to other dates and categories and move records to other subdirectories. Be sure to downgrade or eliminate those tasks that yield little value.

· · · · · · · · · · · · · · · · · · · · · · · · · · · · · · · · · · · · · · · · · ·

Time is one of our most valuable assets. Concentrate on those tasks and activities that help you achieve your goals.

· · · · · · · · · · · · · · · · · · · · · · · · · · · · · · · · · · · · · · · · · ·

## Get Organized—On-Line or Otherwise

When computers became popular in the late 1980s, the media promised that we would create a paperless society. This appears to be true to the extent that we can execute relatively complicated transactions on-line and store lengthy files on floppy disks or a hard drive. Today you can acquire the entire *Encyclopaedia Britannica* on one CD-ROM. In contrast, the version of twenty years ago contains over 25 bound volumes. Even though technology and miniaturization have led to smaller and smaller objects, in many cases we have more clutter, more papers, and more records and files than ever before.

Regardless of how *you* store your information, you'll need to develop a system of organization. Think of a computer hard drive or storage disk as an electronic filing cabinet. Periodically discard or purge all unwanted or outdated files and information. Create subdirectories for all of your important subjects and projects. Many people organize their computer hard drive by customer; for example, for each customer or customer project, they create a new subdirectory. They may even create sub-subdirectories to organize their correspondence, memos, contracts, spreadsheet files and To Do Lists. Design your electronic filing system to match your workflow and help you be more effective.

**Protect your information assets.** Make sure you back up your important files and documents frequently—daily or even hourly. Then secure your backup and other important files in a secure place: perhaps a fireproof safe or a safe off-site location. When you write extremely important and valuable material, make a backup floppy disk of each draft and store the disk at a friend or relative's house. This can be extremely helpful in the event your home or office experiences a casualty such as a fire or flood since it is almost impossible to replace lost information. When you use a network and write an important document, report, manual or letter, copy key drafts—as well as the final printed image—to your company's network server, where they perform regular backups. When you use a PDA, archive the information on both your handheld unit and your PC.

Even when you have a backup on floppy disk, you may find it worthwhile to keep a hard copy printout. You can always use an optical scanner to convert your printed material into electronic files.

• • • • • • • • • • • • • • • • • • • • • • • • • • • • • • • • • • • •
Organize your information—to be more effective and to leapfrog your competition.
• • • • • • • • • • • • • • • • • • • • • • • • • • • • • • • • • • • •

## Valuable Assets

Perhaps the most valuable assets in today's high technology world are information and time. On the one hand, technology gives us the tools to gather and collect infinite information. We can search the World Wide Web ("Web") and locate all sorts of information—including personal details about people and organizations. Despite the end of the Cold War, more and more people join the ranks of cyberspace secret agents every day. This means you *must* keep your personal information and identification as secret as possible. Guard this asset carefully.

The second most valuable asset is time. With technology, many things are at our fingertips. We can buy a wide variety of goods and services with the click of the mouse. We can retrieve up-to-the-minute news flashes on-line. We can even write an e-mail message and program the software to send the e-mail at a designated time. With our wireless phone or PDA, we can send and receive e-mail all the time. With the facsimile ("FAX") machine, we can send and receive written correspondence at the speed of a telephone call.

The drawback with technology is that on many occasions, the pace of activity speeds up too much and technology goads us into responding to others instantaneously. Slow down, take your time and proceed at your own comfortable pace. The other party will wait. Often, people use the speed of technology as a negotiating tactic to encourage you to hurry up and respond or else "they will close the deal with a competitor." This is often a bluff. In many cases, the other party wants your business very badly. Slow down and make them wait! Use technology to your advantage.

## Letters and the Pony Express

In today's world of e-mail, it is relatively easy to respond to an e-mail by clicking the mouse on the "reply" button. Many people find e-mail to be convenient and effective for keeping in touch regularly.

In contrast, a handwritten letter can be a welcome surprise in today's electronic world. When you want to send a special message or make a more personal impression, consider using the traditional paper and pen. Your message may arrive several days later, but the wait and the impact can be very worthwhile.

The traditional mail systems continue to be effective for sending written materials across great distances. In the US, you can send several sheets of paper across the country for the relatively low price of a First Class stamp. A worthwhile use of the postal system is to send hard copies of letters and documents as a follow-up to a FAX. But do factor this in to your calendar: when you use the traditional postal route, it often takes several days for your letter to reach its destination.

. . . . . . . . . . . . . . . . . . . . . . . . . . . . . . . . . . . . .
The mail system remains an effective way to send written material, even if the journey takes longer than e-mail.
. . . . . . . . . . . . . . . . . . . . . . . . . . . . . . . . . . . . .

## Business Cards

In today's fast-paced high technology world, business cards may seem old fashioned. However, business cards continue to offer a handy and convenient way to get in touch with people and help people get in touch with you.

Make sure you print all of your contact points on your business cards including:

- Name.
- Title.
- Company or organization name.
- Business address (don't list your personal address, especially when you use your business card as a luggage tag).
- Telephone number.
- Voice mail (if different).
- Wireless/cell/mobile telephone number.
- FAX number.
- E-mail address.
- Web site.
- Mailing address (if different).

Make sure that people can contact you easily.

You can use the back of your business cards to promote your products and services. Some authors print the titles of their books on the back of their business cards. Many consultants list their services on the back of their cards. Or, you can leave the back of the card blank.

When you leave the back blank, you give people the opportunity to jot down additional information. For example, you might want to write down certain follow-up items or a reminder about something you discussed. Some people use the back of the card to write additional contact points such as a home or wireless telephone number. Other times, you might meet two people from an organization, but receive only one business card; write the other person's name, title and telephone number on the back of the card for your records. You might note another person's name for a referral. When you make notes on the back of your own business card, draw an "X" on the front so you won't inadvertently hand out that card. Another approach is to make it a practice to look at the back side of your card before giving it to someone.

Combine electronic and physical business card systems. Many people use both a business card file and an electronic (software) address book or PDA such as Palm Inc.'s Palm™, Handspring Inc.'s Visor® or Casio's Cassiopeia™ to store names and contact points. For example, you could store all business cards that you use on a regular basis in a card file. You can store seldom-used cards in a rubber-banded stack in your drawer. And you can enter the most active names and relevant data into your electronic system. This can help you gain easy access to many people, especially when you are traveling. If you store these records electronically, make sure you've backed up! Many PDA's have software you can install on your PC for this purpose, so, as long as you ensure your PDA and PC are synchronized, you always have dual copies. As an extra precaution, print your data files on a regular basis and store the hard copy in your files. Many PDA's enable you to scan business cards directly into them, automatically creating an electronic version of someone's business card.

Some people use two or three business card files. You can use one file for cards of people who work in your company. This can be helpful even when your company maintains an on-line employee directory. You can use the other card files for external people. You might want to set up one file or sub-file for customers and one for vendors and suppliers. Organize the business cards in alphabetical order, first by organization and then alphabetically by person. Easy access to people's information helps you get in touch easily. Remember, your business cards are valuable assets—your network—so make sure you protect them.

• • • • • • • • • • • • • • • • • • • • • • • • • • • • • • • •

Business cards continue to be an effective way to keep track of people in your network. And they can provide inexpensive publicity for you. Be generous and give out your cards whenever appropriate.

• • • • • • • • • • • • • • • • • • • • • • • • • • • • • • • •

# Business Communication Across Cultures

Good manners and commonsense indicate that when you are traveling to a different region, you should be familiar with the local customs. There are many good books on this subject, so if you plan this type of trip, we encourage you to learn the cultural practices.

We have some experience with communication practices in Asia, so consider these tips:

*When you conduct business in Japan:*

- When you give someone a business card, use both hands and give it to them face up, so that the writing on the card faces the recipient.

- Never write on someone's business card or immediately put it into your wallet. Hold their card firmly, with both hands. And always study the card carefully and check pronunciation of their name.

- If you place someone's business card in your pocket, place it only in your shirt or jacket pocket that is near your heart. Do not use your pants pocket. If you don't have a suitable pocket, use a pocketbook, purse or briefcase.

- Print your contact information in Kanji on the back of your business cards.

*In China:*

- Never refuse an offer of tea.

*Across Asia:*

- When you enter someone's home, remove your shoes at the door or outside, following your host's lead.

- Greet/Introduce yourself to the most senior person first. Find out whether it is appropriate to bow to more senior people than yourself to show respect.

- When sitting cross-legged, point your feet down to the ground, not towards another person.

*Your Hometown:*

No doubt you have a strong command of the customs and habits in your area. But you might want to review your habits and consider your behavior in light of your audience and the impression you want to make. There are definitely regional differences in communication across the US, so be sure to factor these into your interactions with others.

• • • • • • • • • • • • • • • • • • • • • • • • • • • • • • • •
Be sure to learn, respect and follow the customs of your hosts and hostesses, wherever you are visiting.
• • • • • • • • • • • • • • • • • • • • • • • • • • • • • • • •

## Savvy E-mail Use

Electronic mail or e-mail has become the common method of communication, especially for people across the globe. You can send a message to one person or a larger group of people almost instantaneously....This is an extremely efficient way to communicate with people everywhere.

- Use brief headlines: laser-sharp descriptive titles capture the reader's attention. Many people receive over 100 e-mails every day. When they don't recognize the sender's name or if the e-mail title or subject matter doesn't capture their attention, they will often delete the e-mail. This is also a useful tactic to avoid receiving viruses from the Internet: in most cases, if you don't open an e-mail, any virus the file contains cannot become active on your PC.

- Be as brief as possible. Don't offer more information than is needed. Whenever you can, craft your message in two to four sentences only. For an example of concise reporting, obtain today's edition of *Investor's Business Daily* newspaper: the writers present major news stories in three to five sentences.

- Identify the issue, add the relevant facts, suggest a resolution, offer possible obstacles, present a timetable for response, and ask for agreement. That's it! Many people delete e-mails that exceed three paragraphs in length. When you need to convey a lot of information via e-mail, consider a synopsis or executive summary in the body of the e-mail along with an attached detail file. This way, the recipients can review or print out the attached file if they seek additional information.

- Be friendly; smile as you write; your tone shows in your message!

- Present only brief and concise, non-threatening information in your e-mails. *DON'T be emotional or angry.* Make it a practice to send e-mails that are either factual or complimentary. Unfortunately, once you send an e-mail, you run the risk that the recipient will forward it to other people, including your adversaries. In addition, the recipient can edit your e-mail and then forward it to other people. When you are composing an important e-mail, consider drafting your letter in a word processor such as Microsoft® Word, cutting and pasting your letter into the body of your e-mail. This way, you can spend the time necessary to compose the message you want without incurring significant Internet connect charges. In addition, sometimes the spelling and grammar check tools in a word processor are more powerful than those in an e-mail program.

- Do not use e-mail to send jokes. They're tiresome to read and you can lose your job for sending offensive jokes.

- Separate work and play. Do not use your work e-mail account for personal business. There are countless stories of employees who were dismissed because they sent personal e-mails— messages that criticized their company, discussed starting a new business, or mentioned a new position with another department or another company. Many e-mail systems archive all e-mails for at least two years. Your personal e-mails may live for a long time. Either don't use personal e-mail at all, or maintain your own personal e-mail account and log on at home.

- Be careful about using key words in your e-mails. For example, many banks and financial institutions routinely scan their employees' e-mail for key words such as "guarantee" or "commitment" because these words can have serious legal consequences in the financial world. Read your employee or company manual to learn about your company's policies and visit with your human resources professionals and your information technology specialists on a regular basis.

- Sometimes the best way to keep in contact with people is via e-mail. Many businesspeople travel with their laptop computers and log on to their e-mail wherever they travel around the globe. With the convergence of wireless telephony and data transmission, many people use their wireless telephones and modems to send and receive e-mail.

- Be sure to print out hard copy or backup important e-mails and other electronic files. Due to storage capacity limits, many systems automatically delete (purge) old files. Some e-mail services delete files that are more than one month old.

- Read your e-mail system's user guide. Many systems will log you out when two minutes elapse since you last pressed a key. When this happens, you may lose important information.

- Become familiar with your company's decision-making process. Some companies use the e-mail system to make important decisions while other organizations use e-mail to disseminate information. Learn your company's decision-making process and be careful that you don't behave out of context, for example, asking for decisions or commitments via e-mail if that is not normal in your company.

- Recognize that you cannot evaluate all data when making a decision (with the huge volume of e-mail, there is more information than ever).

- Don't forget to "BCC:" yourself if you want to keep a copy of the e-mail you sent to others. Also, some e-mail systems have the capability to save all copies of your outbound e-mails.

- When you send e-mails to a distribution list (group of people), review the list to make sure that you want to send your e-mail to everyone on that list.

- Don't open e-mails and attached files that you receive from strangers, as they may contain viruses. If you know the sender, request him/her to cut and paste the file into the body of the e-mail.

E-mail operates as a high-speed letter. Be brief and clear, but craft your words carefully.

## Voice-mail Victory

- Deliver brief and to-the-point voice mail ("v-mail") messages. Whenever you can, craft your message in two to four sentences only. Identify the issue and request a specific response. That's it. Consider the following message: "This is John Doe returning your call about the Pinnacle Project. I have scheduled 2:00 to 2:30 PM today for our call. I look forward to speaking with you. My phone number is: 111-222-3456. Please confirm." At other times, you may want to leave a more detailed message that provides information, but don't ramble or take up too much time. People often don't listen to lengthy messages so your final comments could be lost.

- Present only non-threatening information in your v-mails. *DON'T be emotional or angry.* Unfortunately, once you send a v-mail, you run the risk that the recipient will forward it to other people, including your enemies. When you need to compose an important v-mail, you may want to draft a script and then read your script aloud a few times. This way, you will be more likely to actually deliver the message you want and present yourself in the way you desire. Some v-mail systems allow you to erase your message before you deliver it by pressing the asterisk (*) key; other systems do not allow you to erase your message. Spend the time necessary to compose the message using the tone you want to make the impression you desire.

- Do not use v-mail to send jokes.

- Be careful about what you say in your v-mail messages. The most conservative policy is to realize that your employer has the power to review your v-mails. Read your employee or company manual to learn about your company's policies and visit with your human resources professionals and your information technology specialists on a regular basis.

- Sometimes the best way to contact people is via v-mail. Most people who travel dial into their v-mail accounts to check their messages on a regular basis.

- When you want to send an important v-mail and want to keep a copy for yourself, you might want to send the message to yourself as well. With many systems, this is possible.

- Many executives don't check their own v-mail; their assistants review and transcribe appropriate messages. Be sure you are audience-appropriate and clear so you get through the clutter!

· · · · · · · · · · · · · · · · · · · · · · · · · · · · · · · · · · · · · · · ·
Voice mail enables you to communicate with people when they are away from their office or home. As a rule, be brief and clear, while conveying the appropriate tone.
· · · · · · · · · · · · · · · · · · · · · · · · · · · · · · · · · · · · · · · ·

## E-mail and V-mail Responsiveness

- Check your e-mail and v-mail frequently, but don't become an "addict." Technology is a tool and should work for you, not the reverse!

- DO check your e-mail and v-mail while you are on business trips or in training. Your colleagues will thank you...But consider NOT checking while on vacation; we all need a little time away from the office now and again. Most systems enable you to change your e-mail and v-mail messages to state your travel plans and provide alternate contacts during your absence.

- Reply/Reply All/Forward tools are useful but dangerous. Be sure you understand the functionality and be deliberate about your choices.

- "Rule of 3": If a message has been replied/forwarded three or more times, probably the original premise has been diluted. Place a direct telephone call to the individuals involved or arrange a F2F meeting to resolve the issue.

- Urgent/Private/Confidential/BCC: are useful tools. Understand the capabilities of your e-mail/v-mail system (for instance, with many v-mail systems, noone can forward "Private" items, they can only respond to the originator. Also, in some instances, you can designate e-mails: "Don't copy.").

# Security

- Often the best way to protect your information and valuable data is to keep it confidential. Of course, to be effective you'll have to share information. And in today's complex business world, this probably entails sending files electronically across the globe. Always include a confidentiality letter or statement along with your sensitive files, as well as considering encrypting your files.

- Keep your password and/or Personal Identification Number ("PIN") a secret. Once you share your password with others, you lose control over new and previously stored messages (do you really want your colleagues to access the electronic files you have saved with compensation information?).

- DO NOT write down your username and password in the same place! And don't carry your password in your wallet or purse.

- Make your e-mail/v-mail/Web site/network passwords different. Consider this helpful technique: use one core password with a different number in the last place (for example, your e-mail password might be "SUPERWOMAN1", network password "SUPERWOMAN2", etc.). Another alternative, for password systems that are case-sensitive, is to keep your password, but just modify the case (for example, SUPERWOMAN1 and superwoman2). You can also use telephone numbers or another sequence of numbers that you commit to memory. If you are a security professional or if you maintain highly secure information, the above suggestions may not be appropriate for you.

- DO NOT reveal your social security number ("SSN") to anyone unless absolutely necessary. Typically, the only people who need to know your SSN are those organizations relating to income taxes such as the IRS, financial institutions, your employer and sometimes medical providers.

- Always create backup storage systems for your important information and files. Backup systems may consist of floppy disks, off-site storage systems (electronic or paper), photocopies, etc. Protect your valuable information—backup frequently!

- Beware of "Temp Files" in your personal computer. Many software programs create temporary files while you are working on a file. Generally, they are named <filename.TMP>. These Temp Files can be helpful in the event your system crashes; often you can use them to retrieve your work. However, the Temp File can also enable other people to find your work. Also, these files take up valuable disk space. Generally, your system erases the files when you shut down (if your PC is networked, this may be complicated), though sometimes you'll need to go manually delete them.

· · · · · · · · · · · · · · · · · · · · · · · · · · · · · · · · · · · · · · · ·

It is crucial that you guard, care for and protect your valuable information assets. Develop an information protection strategy and follow it consistently.

· · · · · · · · · · · · · · · · · · · · · · · · · · · · · · · · · · · · · · · ·

# The Paperless Society

- Keep your systems and databases in order—your e-mail filing system may be more important than paper files if your company uses e-mail extensively to share documents, contracts, letters, strategic plans, presentations, and other important documents and information.

- Many e-mail systems contain "filters" that you can setup to screen out less-than-critical messages to prevent "information overload." You don't need to read and respond to every e-mail!

- **Travel Tactic**—Create two "Travel" folders. In one folder or subdirectory, save all of your travel profile information such as important contact numbers and your frequent flyer data. In the second travel folder, save everything for an upcoming specific trip. In one place, you can keep the itinerary, reservation numbers, directions, presentations, meeting schedule, customer information, etc. In many companies, you can make your travel reservations through e-mail (e-mail reservations are often more efficient than waiting on the phone while the travel agent checks availability and makes the reservations). You can also cut and paste from the e-mails into an e-calendar, so others can actually locate you! Print both folders before you leave your office. Then delete the trip file when you submit your expense report for that trip, so your folders remain as clean as possible.

- When sending e-mail attachments, be sure to identify the filename so the recipients can find your attachment. Identification becomes particularly important when you attach multiple documents– when you do this, give the filename and a small explanation of the contents (e.g., "Dear Michael, I've attached <Cat-budget.xls.> This file contains the budget and spending estimates for Project Cat. Please review the data to make sure we included all of the relevant information."). Many people will not open files attached to an e-mail in fear of activating a virus. To avoid sending attachments, you can cut and paste the contents of the file directly into the e-mail, send a floppy disk to the recipient or perform a virus scan on the file and encrypt it before sending.

- When you send relatively large attachments, especially files that contain graphics or presentations such as Microsoft® PowerPoint® documents, ZIP the files to make them smaller. It's a good idea to get into practice of *ZIPping* all large documents, so you save hard disk space on your PC. There are great tools for this, available free for basic versions—you can download the files from the Internet. Two examples are WinZip® (http://www.winzip.com) and Netzip® (http://www.netzip.com).

- You can use the e-mail system to notify you that your recipients received and opened your e-mail. Print hard copies of important electronic files. This can be helpful to create an e-trail and to log your involvement in a project.

• • • • • • • • • • • • • • • • • • • • • • • • • • • • • • • • • • • •

In the "paperless" society, organizing your information is more crucial than ever before. Without papers, there's nothing to shuffle on your desk!

• • • • • • • • • • • • • • • • • • • • • • • • • • • • • • • • • • • •

## Personal E-mail

- *Corporate e-mail and v-mail are corporate property, neither personal nor private property.* This means your company can require you to save, print or otherwise provide any communication. The company has the right to retrieve any communication from the archives whether or not you agree. Many e-mail systems automatically archive e-mail files for at least two years.

- Get a personal account if you plan to conduct personal interactions via e-mail—to look for a new job or start your own business. E-mail addresses are available for free from some vendors:

    Yahoo: http://www.yahoo.com

    Netzero: http://www.netzero.com

    Juno: http://www.juno.com

    Typically you need only a web browser to access these free e-mail offerings. Many public libraries offer PC's if you don't personally own one and, in many locations, there are "Internet cafes" where you can rent time on a PC connected to the Internet.

- *Don't send jokes, chain letters or derogatory comments via e-mail or v-mail.*

. . . . . . . . . . . . . . . . . . . . . . . . . . . . . . . . .
Don't send personal e-mail from your corporate e-mail address. The wiser strategy is to maintain a personal e-mail account and log on outside your office.
. . . . . . . . . . . . . . . . . . . . . . . . . . . . . . . . .

## E-mail Do's and Don'ts

- **Beware of protocol:** use capital letters for a strong emphasis or to express anger. Underlines or asterisks (*) on either side of a word are a friendlier way to emphasize a point without alienating your reader.

- *Always use* the spell-check and grammar-check tools. Be sure to proofread your writing even when you use the software tools. The tools might miss a word as follows: suppose you intended to write "ratios" but accidentally typed "rations". Since "rations" is spelled correctly, the software is likely to bypass it. Use capital letters when appropriate.

- Use a signature file or other identifier at the bottom of your e-mail to provide phone/FAX contact information and, if a corporate e-mail account, your corporate affiliation. Often you can incorporate this information into a header or footer.

- If you are sending e-mail from your business account, you might consider adding a disclaimer such as "The opinions expressed herein are my own and do not represent the [XYZ] Corporation."

E-mail is a system for written communication. Like all other systems, develop your own e-mail strategy and set of rules. And follow those rules consistently.

## Staying in Touch

- E-mail and v-mail are fabulous vehicles for staying in touch, in a world where none of us has enough time to see family, friends and acquaintances. How many times would a simple, "I'm thinking of you" brighten your day? Parents and grandparents just adore this. To delight grandma, e-mail a scanned picture of your child's soccer game to her. Many people create personal Web sites specifically to broadcast digital pictures of their newborns: one friend continues to maintain these initial photos three years later!

- E-mail is also an incredibly inexpensive way to stay in touch with people who live far away, or who travel often. Magazine articles sometimes feature couples who embark on year-long sailing trips and check in with friends through periodic e-mails—how else would the world know about the local entertainment in Havana? The Reply button makes it relatively easy to respond quickly. And wireless telephones help you connect from remote locations.

Use communication devices and systems to keep in touch with your family, friends, contacts and colleagues.

## Conference Calls

- Here are some basic do's and don'ts: to avoid interruption, setup conference calls ("concalls") for longer duration than you need and for more dial-in lines than you anticipate. Publish the telephone number and passcode when you first setup the meeting, not in later e-mails. Distribute the meeting materials (documents, agendas or presentations) as far in advance as practical so the audience can fully review the material and prepare for the meeting; one or two days before the conference call is optimal.

- General protocol is to take a roll call at the beginning of the call. During the call, it is useful for each participant to identify him/herself when making a statement. This provides context to the other listeners, a very helpful thing when you can't see others in cyberspace.

- If your concall is a meeting, request the meeting vendor to identify folks when they enter and exit the call so the remaining participants know who is on the line.

- If your concall is a presentation or training session, here are some options to request of the meeting vendor: attendee names and affiliations, tape the session for possible later playback, have the operator open the lines for the speakers only during the presentation and then control the attendee lines for the Q&A session. Many companies now hold global conference calls to present important information to groups of people all over the world at the same time. In the investment community, companies regularly hold calls to present their earnings, where investors use the dial-in number to listen to the call. Technology helps more people gain access to information.

- Some concall vendors you might use include:

    Qwest Communications (http://www.qwest.com or 800.860.8000).

    Global Crossing (http://www.globalcrossing.com or 888.825.5861).

    AT&T (http://www.att.com or 800.222.0400).

    MCIWorldcom (http://www.wcom.com or 800.MCI.WCOM).

- With the new Web-based meeting services like Contigo (http://www.contigo.com), you are also able to ask survey questions and receive instant feedback from your audience. One tactic: keep each survey question and its answers brief. It is easier and more reliable to get feedback when the options are multiple choice with just a few choices (for example, "yes" and "no;" or a few price points "under $100," "$100–$250," "over $250").

- You could also try Microsoft® NetMeeting® (http://www.microsoft.com) with a group of friends to have a group chat, or one of the Internet-to-Phone vendors (such as http://www.net2phone.com or http://www.dialpad.com) for inexpensive Internet phone calls. This technology is relatively new and has sometimes inconsistent performance, but will undergo rapid improvement over the next few years—they may also begin charging higher rates, so you'll want to keep checking for updates.

- For people in the same location, you may want to reserve a conference room. This gives groups of people the opportunity to exchange notes and observe facial expressions.

- If you tape the call, you are legally obliged to disclose this information at the beginning of the call and offer participants the ability to opt out of the call.

- Use the Mute button carefully. Many conference call participants press the Mute button to make side comments. Make sure the telephone is actually on "Mute" before you speak. Many people reveal confidential information accidentally.

- At the end of a concall, be sure to make a fresh call to wrap-up with colleagues—you want to ensure that your private remarks remain private and that you are off the line from the other parties.

· · · · · · · · · · · · · · · · · · · · · · · · · · · · · · · · · · ·
Be careful and deliberate when you speak on a concall. You never know who is listening and whether someone is recording the conversation.
· · · · · · · · · · · · · · · · · · · · · · · · · · · · · · · · · · ·

# Nothing Beats Face-to-face

- You can accomplish many things in today's high-tech, fast-paced electronic world, but the best way to build a relationship is in person, one-to-one, one step at a time. Make sure you pay attention to our very human need to see and talk to other people, especially when technology can make communication so impersonal.

- When you meet with someone F2F, focus on that meeting. Don't field telephone calls or process e-mail during the meeting. There's enough time later. The limited face time you get with someone is valuable, so don't waste it. Don't cheapen it or be rude—let your v-mail and e-mail tools accumulate your messages; they will wait until you finish the meeting.

- In our increasingly electronic age, letters, "thank you" notes and holiday/birthday cards are critical. Many people send electronic versions of these notes–but they don't seem to be a satisfying substitute for handwritten cards, especially when your computer doesn't contain the browser software plug-in required to view the card! For business, recipients usually appreciate personal written notes, especially when you're in sales or customer service.

## Pager Etiquette

- Paging technology and coverage where you can receive a page across the US is of relatively high quality today. You can select coverage ranging from your local area, US-wide, worldwide and even airborne during a flight. Technology options include simple digital (numbers only) or alphanumeric (text messages, often with a maximum length) pagers. And the price of the service has dropped too: a number of service providers offer US-wide coverage on a digital pager for $30 per month, with coverage nearly everywhere!

- Also, you can obtain guaranteed service, where the page will continue to be sent until you acknowledge that you received it. Another option is two-way paging, where you can receive and send alphanumeric (text) responses (typically with a maximum length). Most services include a number of generic messages such as "Call me" or "On my way." Or you can type your own message.

- Many paging services can send you news, sports headlines and stock market updates.

- If you need to reach someone immediately, you can type "911" in front of a phone number or in the beginning of a subject line, so the recipient will respond to your page immediately. Different people define "emergency" in different ways so use this tactic carefully. Digital pagers can handle 14 digits, so you can send "1 911 202 555 1212."

- *Send short text messages only* when you use alphanumeric.

- Wait for your recipient to respond! Remain by your phone and don't keep paging, as it often takes 10 to 20 minutes for the page to reach your recipient.

- Always consider the social aspects of paging technology. Pagers are tools to help you reach other people and enable others to reach you. You might want to turn your pager to "vibrate" when you're wearing it, when you are in an important meeting or in a quiet room or restaurant. Complete your current conversation or the meeting before you return the page, and excuse yourself to another location to return the page if you need to use the telephone. Since F2F meetings are relatively scarce today, use this precious time wisely. Be sure to turn off your pager when you are in an airplane.

- Some paging vendors you might use include:

    Pagenet (http://www.pagenet.com or 800.PAGENET).

    AT&T (http://www.att.com or 800.222.0400).

    MCIWorldcom (http://www.wcom.com or 800.MCI.WCOM).

## Legally Binding FAXes and E-Signatures

- In many states and for many business and personal transactions, the courts now consider FAXes to be legally binding. When you negotiate a transaction, be sure to consult your attorney about the legal status of a FAX. If your materials, your transaction, or the business you are conducting is important, *send or request a hard copy via regular mail.* Also, be sure to use a DRAFT stamp—this means that the current version of the document is not the final document. If your material is important, consider sending a hard copy of the FAXed material and mark the top of the first page "follow up to FAX." Always protect your interests, even when you have to take more time to consult your legal advisers.

- Today you can acquire a number of free or inexpensive FAX software products. With computer software and different system configurations, you can FAX a document from your PC to the recipient's FAX machine or even to their PC: this applies to documents, presentations, graphics, etc. While you can use e-mail, today most people have access to a FAX machine. Most copy centers including Kinko's® have FAX services: the cost is approximately $2 to $4 for the first page and $1 per page for each subsequent page. When you send and receive FAXes frequently, you may want to sign up for a personal FAX number on the internet or purchase a basic FAX machine for your home; the cost is approximately $150 and often, these machines can make photocopies. Another traditional but effective method is to send your materials via certified or registered mail through the US Postal Service.

- Mark your FAX cover sheet *"Confidential. For [Recipient] Only."* Another helpful tactic is to call the recipient when you send a FAX and ask him/her to wait at the receiving FAX machine to receive the FAX as soon as it arrives. This minimizes the likelihood that strangers or competitors will obtain your material. This is especially worthwhile at a convention, hotel or other location where strangers roam.

- **Travel Tactic:** Many times, the FAX is a useful alternative to e-mail for reviewing documents when you travel. The FAX makes it easier to proofread documents than to read them on-screen; and many modems from hotels and other remote locations are relatively slow, especially when you download a lengthy document or presentation.

- **Times to FAX:** Often, the optimal times to send FAXes are at 10:30am and 1:30pm in the recipient's time zone. This gives the recipient the opportunity to clear and sort FAXes received overnight and during the lunch hour. Some of the more sophisticated FAX machines enable you to store a FAX in memory and pre-program the machine to send your FAX at a specified time.

- **Digital or E-Signatures:** In June 2000, the US Government passed legislation that signatures sent over the internet are equivalent to signatures on a piece of paper or a contract. In other words, your electronic or e-signature can be legally binding. Consumers must elect to join or accept e-signature documents and must authorize vendors to send notices, records and documents electronically rather than on paper. As with most new acts and policies, this bill will have a great impact on business and e-commerce. Make sure you work with your lawyer to understand the specific implications to your business and contact each vendor you deal with to learn how they implement e-signatures and the consequences to you and your organization.

## Some Helpful Wireless Phone Facts

- Most wireless phone conversations are NOT secure. You probably know people whose wireless phone numbers have been stolen. Today, most service providers have software in place to detect taps and prevent your number from being compromised. Despite the power of these services, an intruder can still eavesdrop on your call. Unfortunately, even a law-abiding citizen may mistakenly and unintentionally pick up your call. Keep this in mind when you conduct confidential conversations. In highly competitive businesses and sensitive situations, you may want to mention to the other party that you are using a wireless phone so the other participants can be mindful of the confidentiality issues. Consider using codenames when you use a wireless phone; you can use a codename for a person's name or to designate a project name (for example, "Project Chicago" or "Mr. B").

- Caller ID is a service that writes the caller's telephone number in your wireless phone's display screen. The major carriers provide this fee-based service.

- In most states, your telephone number is available to people you call unless you specifically block your number from displaying. In most states, your carrier must offer you the ability to block your number at your request; blocking your outgoing number is a free service, but you must request it. You may have to make several requests to successfully block your number and there may be quite a delay from the time of your request until blocking.

- **Wireless Phone Privacy Limitations and Risks!** When you use your wireless phone in a small space like an airport or conference center, someone may hear you—so be careful what you say.

Remember, people can poach your wireless phone number, and this is easier to do in areas of high wireless phone concentration. Whenever practical, use landlines in airports and at conferences for the above reasons and because, with all the background noise, the wireless phone may not be particularly effective.

- Transmission quality also varies by region. You may have to use a landline in many locations because there are often a limited numbers of cells available, resulting in busy signals.

- The lesson: don't count on your wireless phone to work at airports or convention centers or other areas of extremely high traffic. In many areas, service is highly variable. Often a landline works, or stepping away from a crowded venue increases your chances for success.

- Some wireless vendors you might use include:

    AT&T (http://www.att.com or 800.222.0400).

    CELLULARONE (http://www.cellone.com or 800.CALLSWB).

    Omnipoint (http://www.omnipoint.com or 800.RU4.OMNI).

    Sprint PCS (http://www.sprintpcs.com or 800.253.1315).

- If you are a heavy user of wireless phones, you might want to get a headset or speaker.

# PDA'S and Electronic Organizers

- PDA's (Personal Digital Assistants) enable you to take notes, maintain To Do Lists, centralize your telephone and address directory, keep your schedule, make mathematical and financial calculations and play electronic games. Perhaps the most helpful feature of the PDA is that you can condense 5,000 slips of scrap paper into one electronic file. PDA's offer extremely efficient data portability: you can keep your up-to-date schedule with you, as well as change it on demand. Most PDA's have a keyboard and a stylus (plastic pen) so you can touch the screen and compose your material. So far, PDA's are also smaller than other planners and easier to use. You can use a PDA to remind yourself of the topics for a meeting, or your shopping list–and to tell you when to do these things.

- Wireless modems often come with the new generation of PDA's so you can send and receive e-mail. Many new models also offer infrared exchange of information among each other.

- Palm™ is a very popular brand today. There are other PDA's available from Handspring, Hewlett-Packard, Sharp and Casio. There remains significant corporate investment in this space so tomorrow's machines are likely to be much more sophisticated. In today's market, the decision appears to be between the Palm or Visor PDA with their "Graffiti" handwritten interface on the handheld, or a PDA with a keyboard, like the Sharp machines. Simplicity is key to your choice: most people look for a unit that does most of the functions they want, is easy to use, small and has a long battery life.

- **Backup Your PDA data** at least twice a day and whenever you make significant updates. The machines themselves are relatively fragile and when you back up your data to your PC, you save your long-accumulated and important data. Palm calls its back-up feature HotSync®. With many PDA's, you can run dual systems: one on your handheld unit and the second on your PC. At work, you can use the internal corporate network to update both the data in your handheld unit and the PC; you can also run back-up systems on your home and office computers. Use different PDA interface cradles and shuttle the data between each system so you keep all of your systems up to date.

- Organize your PDA in the same way you would organize a paper filing system or your PC. You can create subdirectories for particular subjects and projects.

- Print your data and store the hard copy in your files. PDA's allow you to print your data. You may find it helpful to print your telephone directories and To Do Lists, especially when you travel regularly.

- **Monitor your battery indicator closely.** Many PDA systems contain a back-up battery, but you have only two or three minutes to install new batteries when you remove the old batteries. If you exceed the time limit, you risk losing your data.

· · · · · · · · · · · · · · · · · · · · · · · · · · · · · · · · · · ·

PDA's put your critical information at your fingertips. With a combination PDA and wireless telephone, you can access the Web for a variety of services.

· · · · · · · · · · · · · · · · · · · · · · · · · · · · · · · · · · ·

## The Future is Here Now—Well Almost

- **Mobile e-mail**. With the new wireless handheld devices, you can send and receive e-mail nearly anywhere.

  BlackBerry™ currently appears to be the most popular of these gadgets. Designed to be on all the time so you can send and receive on-demand, these products include an embedded modem and keyboard and integrate with most e-mail systems. This device also serves as an organizer. BlackBerry™ is offered by Research In Motion Limited (http://www.rim.net).

  You can expect more development in this space.

- Another service you should be aware of: single number access with "find/follow me" capability. This is a relatively simple concept, but difficult to implement. You, the subscriber, sign up for one single number. That's right, one number for your phone/FAX/wireless. Callers hear, "Please state your name" and the phone agent rings you so you can decide whether to field the call or receive a message. For this to be effective, you must continually inform the system agent with your whereabouts so the network can locate you. And you need to answer the call from the phone agent, or your callers will go to v-mail. This technology can be useful for those who are disciplined enough to use it and it is reasonably mature right now. Some people don't like talking to the machine voice agent, but you may not mind it.

- Some vendors you might use—offering single number access for phone and FAX with "find/follow me" capability—include:

    888mynumber.com (http://www.mynumber.com).

    AT&T (http://www.att.com or 800.222.0400).

    uReach.com (http://www.ureach.com) - v-mail only; does not have "find/follow me" capability.

    Wildfire® Communications (http://www.wildfire.com or 781.778.1500); this service also permits subscriber to return calls.

- **Convergence combines e-mail and v-mail.** This new service is available, but not yet universally. Convergence means you'll have one "inbox" for all types of messages: v-mail, e-mail and FAX. All this data will be available on your screen. Click on one message and it displays or reads to you, as you choose (so you could read a v-mail or hear an e-mail). Checking these multiple systems—work, home and wireless—and checking them each for updates, can be very time-consuming.

- Some vendors you might use—all include free single number access from phone or web for v-mail and e-mail—include:

    myTalk.com (http://www.mytalk.com).

    Onebox.com (http://www.onebox.com)—also includes FAX.

# One Final Word...For the Moment

When Alexander Graham Bell invented the telephone in the 19th century, he revolutionized communication technology. As great a leap as that was, the pace of technological advancement and the rate of change have accelerated since then. The good news about current technology is we have almost infinite information at our finger tips, especially with the Internet and its many search engines. For better or worse, technology enables us to find people and others to find us 24/7—invades our privacy.

Technology gives you power and can help you become much more efficient and effective on the job and in your life. Take your time! Plan a strategy regarding how you can best use today's technology in your life. Remember—regardless of the current state of technology and which generation of technology you use and its level of sophistication, *the content of your message still rules.*

Rules help you chart a path of progress and help you attain your goals and objectives. It seems that life's pace quickens all the time. One classmate of ours recently discarded a laptop PC from 1988—you can imagine how slow the processor was and the weight of the machine. That technology is definitely outmoded now! With today's computer and telecommunications technologies, we can process and transmit information at the speed of a telephone call. And, with optical technology, we will approach the speed of light. But once again, regardless of the method of communication, the fundamentals of your message remain the same. The message is the most important aspect of your communication. To succeed you must develop rules, and apply those rules consistently.

*Follow these steps to communicate through technology and take control of your life:*

- Think about what message you want to send.

- Consider the tone of your message.

- Determine the optimal time to send your message.

- Determine the best way to send that message.

- Decide who should receive your message.

- Protect your intellectual property.

- Carefully monitor your choice of tone and language.

- Consider the legal and ethical implications of your actions.

- Choose your technology carefully in light of privacy and confidentiality issues.

- Be as brief as possible!

- Learn to speak and write well. Articulate your thoughts and ideas clearly.

- Perhaps most important to success, *proceed on your own timetable.*

Be on the lookout for new technologies and new upgrades to electronic devices. At the same time, be cautious about your purchases since compatibility with your existing tools and ease of use of new devices continue to present important issues. And remember...technological obsolescence occurs rapidly and our needs for additional memory and storage space continue to expand.

Best wishes for success in developing and enhancing your communication. Establish a set of communication rules and follow them. Take control of your career, business and your life today.

You can communicate with us via letter or e-mail. We will respond.

Eve Milrod and Arthur VanDam

C/O Career Advancement Center, Inc.
PO Box 436
Woodmere, NY 11598-0436 USA
E-mail: caradvctr@aol.com

---

*Special Offer:* Order our new book *Building a Million Dollar Service Business*. This valuable guide shows you how to start a more profitable venture and grow your business smartly: $19.95 per copy plus $4.00 shipping and handling. To order your copy, call today: 1-800-295-1325. Better yet, send a check or US postal money order for $19.95 per copy and you save the shipping and handling charge.

# How to Read Legal Documents and Make Better Business Deals

By Joseph Gelb
(A new publication from this publisher)

This booklet is designed to help you build your skills in reading legal documents, gain an understanding of the basic premise behind legal documents and contracts, protect yourself and make better business deals for yourself and your company.

In our small business and legal practice, we studied our client base and reviewed the outcomes of numerous cases, mergers and acquisitions, wills, trusts and estates, and investment transactions.

In almost all our dealings, the best outcomes arose because the parties had a clear set of goals and objectives and had *actually read* the legal documents. Most of our clients are *not* attorneys and the majority lacks a legal background. Those people who took the time to read and study the legal documents had a clearer understanding of what they wanted to achieve from a given transaction and what they were willing to give up in exchange. In effect, they were in a stronger position to make more informed decisions.

When you gain an understanding of the nature of the documents you are about to sign and at the minimum, develop a perspective for what makes sense and what doesn't, then you are more likely to protect yourself, and achieve a greater percentage of your goals and objectives. Here's to more profitable and beneficial transactions and business dealings.

For more information on this publication, see page 52.

# THE CAREER ADVANCEMENT CENTER CATALOG

## Mastering Communication Through Technology

By Eve Milrod and Arthur VanDam

54 pages; $10.00

## How to Read Legal Documents and Make Better Business Deals

By Joseph Gelb

This valuable booklet will help you decipher legal documents. Explains the concepts behind contracts, agreements and documents, and common clauses. Full of useful tips on what to look for (including pitfalls and traps), negotiating better terms and conditions, analyzing what the language means and how to protect yourself. Tactics to improve your critical reading, thinking and business skills.

16 pages; $5.00

## Book Promotion Made Easy—Event Planning, Presentation Skill and Product Marketing

by Eric Gelb

Increase your profits by enhancing your presentation skills and mastering the art of personal selling. This book will show you—step-by-step—how to develop a marketing strategy; promote yourself and your business; and merchandise your product offering.

54 pages; $12.00

## Self-Publishing Strategies Vol. 1—Getting Started

by Eric Gelb

How you can self-publish your next book, special report or article. Friendly discussion in Q&A format. Self-published author Eric Gelb explains how you can earn more money with your next publication. Helps you evaluate the process and create an action plan. Ideal for businesspeople who want to promote their business and writers who want to break into this exciting and lucrative field. Useful information that will help you succeed.

32 pages; $12.00

## Personal Budget Planner—A Guide for Financial Success

by Eric Gelb

Unusual approach to financial planning and money management that blends anecdotes with solid examples. Plenty of concrete tips and examples for avoiding common problems and getting ahead in the financial world. Financial tables encourage calculations and know-how rather than vague speculations and theories. Recommended in Inc. magazine and Business Week.

106 pages; $19.95

## Checkbook Management—A Guide to Saving Money

by Eric Gelb

Discover how to use your checkbook to save more money and build more wealth. Shows how to reduce your checking account and ATM fees. Illustrates how to give yourself a financial check-up by analyzing your checkbook to isolate signs of trouble. Transform your checkbook into a powerful financial planning tool. Commonsense financial guide.

96 pages; $6.50

## Budgeting At Your Finger Tips

by Eric Gelb

Nuts and bolts booklet designed to help you budget your money and organize your finances. Explains every day financial concepts. Contains user-friendly worksheet with line-by-line instructions.

12 pages; $3.50

## Tax Accounting for Small Business

by Joseph Gelb

How to reduce your small business taxes today. Essential tool for the small businessperson. Handy guide shows you how to prepare IRS Form 1040 Schedule C. Hundreds of easy-to-follow tips on how to structure your business and keep accurate records for revenue and expenses. Explains income taxes in plain English.

128 pages; $14.95.

## Building A Million Dollar Service Business—Taking Your Business Idea from a Dream to a Profit

by Joseph Gelb

Extremely useful guide to help the prospective entrepreneur navigate the basic steps of starting a business. Includes corporate form, registration requirements, accounting and bookkeeping, choice of business year, taxes, cash planning, forecasting and financing. Includes the more common business tax forms.

128 pages; $19.95

## The Small Business Advisor® newsletter

published by Joseph Gelb

The newsletter for the small businessperson—jam-packed with useful and timely information. Provides practical advice and useful information on starting and managing a small business. How to grow your business sensibly and profitably. Topics include finance and cash flow, human resources, information systems, insurance, law, management, marketing, operations and tax. Presented in an informative, implementation-oriented style. Each monthly issue contains 16 pages.

12 monthly issues mailed within US: $45.00

24 monthly issues mailed within US: $80.00

Foreign Airmail, add $45.00/year

• • • • • • • • • • • • • • • • • • • • • • • • • • • • • • • • • • • • •

Career Advancement Center books and publications are designed to help you build wealth, advance your career and get ahead in today's world. Why not order now?

• • • • • • • • • • • • • • • • • • • • • • • • • • • • • • • • • • • • •

## Career Advancement Center, Inc. Order Form

☎ For more information, call (516) 374-1387 Monday through Saturday: 9:00 am to 6:00 pm EST. For faster service, call our ordering center at (800) 295-1325. Via FAX (516) 374-1175.

www Internet: smallbusinessadvice.com.

✉ By Mail: PO Box 436, Woodmere, NY 11598-0436 USA.

Checks or US Postal Money Orders; MasterCard or Visa only.

| Title | Quantity | Price | Total |
|---|---|---|---|
|  |  |  |  |
|  |  |  |  |
|  |  |  |  |
|  |  |  |  |
|  |  |  |  |
|  |  |  |  |
|  |  |  |  |
| **Subtotal:** |  |  |  |
| Add New York State and New York City Sales Tax if applicable (on all items except newsletter): |  |  |  |
| **Total Order Price:** |  |  |  |
| Shipping & Handling: $4.00; for quantity orders, shipping charge will be based on the United Parcel Service ground shipping charges. No extra shipping charge applies to the Small Business Advisor® newsletter. |  |  |  |
| **Total Amount Due:** |  |  |  |

Please specify different address if billing and shipping addresses are different.

Name _____

Complete Address _____

Telephone #/e-mail _____

Payment Type (please check one):

❏ Check/US Postal Money Order     ❏ MasterCard     ❏ Visa

Credit Card # _____ Exp. Date _____

Cardholder signature: _____

Please allow 4–6 weeks for delivery. Rush delivery available. Thank you for your order.